YOUR KNOWLEDGE HAS VALUE

- We will publish your bachelor's and
  master's thesis, essays and papers

- Your own eBook and book -
  sold worldwide in all relevant shops

- Earn money with each sale

Upload your text at www.GRIN.com
and publish for free

Bibliographic information published by the German National Library:

The German National Library lists this publication in the National Bibliography; detailed bibliographic data are available on the Internet at http://dnb.dnb.de .

Imprint:

Copyright © 2019 GRIN Verlag
Print and binding: Books on Demand GmbH, Norderstedt Germany
ISBN: 9783668920958

This book at GRIN:

https://www.grin.com/document/460654

Rangappa Yaraddi, Annesh G. M.

The Impact of Integrated Farming System on Farmer's Agricultural Income

A Case on Davangere Dist, Karnataka State

GRIN Verlag

"The Impact of Integrated farming system on farmer's agricultural Income"

-A case on Davangere Dist, Karnataka State.

Rangappa Yaraddi

Karnataka State Rural Development and Panchyat Raj University Gadag, Karnataka, India.

Annesh G M

Karnataka State Rural Development and Panchyat Raj University Gadag, Karnataka, India.

ABSTRACT

In this study the main focus was give to study the influence of integrated farming on farmer's income in Davangere district of karnataka state. The integrated farming is an ancient method used all over the world to reap the different benefits from agriculture land. In India Integrated farming system was been in use since the early civilization and settlements. In recent days the integrated farming is becoming more popular due to its characteristics and properties.

In this study around 39 farmers are consulted to collect the primary data. The primary data was collect using pre administered questionnaire. The questionnaire was consisting of ten different variables related to familiarization and farmers income of integrated farming in Davangere district of karnataka state. The collected primary data was analyzed using the tools like MS-excel and SPSS. In this study Z-test was used to confirm the hypothesis along with cumulative frequencies method. The significance level is taken as five percent.

The study was concluded with the outcome of "there is a significant positive influence of integrated farming on farmer's agriculture income". During the study it was also observed that, the availability of water throughout the year is a key variable for farmers yield and income.

Keywords: Integrated farming, farmers income, significance of integrated farming, agricultural income etc.

INTRODUCTION

In recent years, food security, livelihood security,water security as well as natural resources conservation and environment protection has emerged as major issues worldwide,Developing countries are struggling to deal with these major burning issues .Globalization has been accepted by everyone across the globe that sustainable development is the only way to promote rational utilization of resources and environmental protection without hampering economic development. Developing countries around the world including India are promoting sustainable development through sustainable agricultural practices which will help them in addressing socioeconomic and environmental issues simultaneously. Integrated farming method holds special importance in this context. In IFS system none of the byproducts are wasted. Byproduct of one system becomes the inputs for other crops in an integrated farming approach as compared to the existing mono culture approaches. IFS has multiple advantages in the areas such as sustainability, food security, farmer security and poverty reduction etc.

LITERATURE REVIEW

Article title "Well designed integrated farming system can help double Farmers increase by 2022" by Soleplates simoom on 20 March 2018 in Down to earth journal, lists about the"well designed" integrated farming system (IFS). The study reveals the vision of the government of India double Farmers income by 2022 by IFS. As per the study the IFS enhance the farmer's income and IFS can act as a instrument for rural development, agriculture with animals husbandery and other occupations related to core agriculture practices. the study gives the statistics of Tripura Farmers who earns 1.6 million from his 2.6 Hector land by IFS. As per the author of this study the IFS can be a powerful tool to double farmers income and improve their lives. The IFS will increase the production, making effective base of inputs cost, reducing post harvesting loses; make value addition and reforms in Agri marketing which will make double income for the farmers.

Article title" Scientifically designed integrated farming systems protected to boost farming income" by capital markets on 02 August 2018 in business standard journal. Lists about the Scientifically designed and tailor -make IFS.The study focus about the institutional integrated system, the intentional integrated system which address the many objective of increased production, profit cost reduction through recycling, family nutrition sustainability, ecological generation etc, based on the past study of crop production and nutritional security the government support location specific integration of field crops.

Article titled " Sustainable integrated agriculture and rural development policy" in "Advances in crop science and technology" on 18 Nov 2015 author Hari Prasad Silwal lists the mechanizes mass scale commercial agriculture system, the study gives the collaborative/ organizational structure institutionalized organized uses group amongst farmers, improved farm mechanizes with modern technologies and activities.

OBJECTIVES

1. To study the effect of integrated farming system on farmers agricultural income.

HYPOTHESIS OF THE STUDY

H1: Integrated farming system has no effect on farmers agriculture income.

H0: Integrated farming system has a positive effect on farmers agriculture income.

METHODOLOGY AND DATA COLLECTION

During the study, primary data was collected through structured questionnaire to gain insight about the various aspects of integrated farming system from the farmers. For the study secondary data was collected from various articles, websites and research publications. The sampling technique used is non probability sampling because specific sample have been selected as per convenience of the researcher.

Primary data: The relevant data collection instrument (Questionnaire) was designed and self administered during the collection of primary data in Davangere district of karnataka state.

Questionnaire: A questioner consisting of 10 items including the factors like 'agricultural income of farmers', 'impact of integrated farming on farmers income' etc   was self administered and filled during the interaction with the respondents in Davangere district of karnataka state.

Description of the data collection tool:
Part I- Personal details of the respondents.
Part II- Impact of integrated farming system on farmers income.

Secondary data: secondary data was collected from various articles, websites and research publications, books etc.

Sample Point: The primary data will be collected from the farmers who are practicing integrated farming in their field.Around 39 samples are collected from integrated practicing farmers in Davangere district.

Sampling Method: Simple random sampling method was adopted for the primary data collection.

Data Analysis: data analysis tool such as MS-Excel and SPSS was used during the study. Z-Test and cumulative frequency methods are adopted to verify the data statistically.

ANALYSIS OF DATA

Tabulation of the Responses:

Table 1.1: Frequency distribution of the response for the statement **"I am aware of Integrated farming methods & uses and I have used at least once in my farm land."** (V1).

V1

		Frequency	Percent	Valid Percent	Cumulative Percent
Valid	SD	2	5.1	5.1	5.1
	D	1	2.6	2.6	7.7
	N	2	5.1	5.1	12.8
	A	30	76.9	76.9	89.7
	SA	4	10.3	10.3	100.0
	Total	39	100.0	100.0	

Table 1.2: Frequency distribution of the response for the statement " **I am happy with the performance of Integrated farming methods in my agriculture farm land** " (V2).

V2

		Frequency	Percent	Valid Percent	Cumulative Percent
Valid	SD	1	2.6	2.6	2.6
	D	1	2.6	2.6	5.1
	N	3	7.7	7.7	12.8
	A	24	61.5	61.5	74.4
	SA	10	25.6	25.6	100.0
	Total	39	100.0	100.0	

Table 1.3: Frequency distribution of the response for the statement " **I am comfortable with adoption of Integrated farming methods in my agricultural field** " (V3).

V3

		Frequency	Percent	Valid Percent	Cumulative Percent
Valid	SD	1	2.6	2.6	2.6
	D	3	7.7	7.7	10.3
	N	4	10.3	10.3	20.5
	A	20	51.3	51.3	71.8
	SA	11	28.2	28.2	100.0
	Total	39	100.0	100.0	

Table 1.4: Frequency distribution of the response for the statement "**Due to the adoption of Integrated farming methods in my agricultural field, I avoid the usage of inorganic fertilizers**"(V4).

V4

		Frequency	Percent	Valid Percent	Cumulative Percent
Valid	SD	1	2.6	2.6	2.6
	D	2	5.1	5.1	7.7
	N	16	41.0	41.0	48.7
	A	15	38.5	38.5	87.2
	SA	5	12.8	12.8	100.0
	Total	39	100.0	100.0	

Table 1.5: Frequency distribution of the response for the statement "**Adoption of Integrated farming methods has brought down the fertilizer cost in my agricultural field** " (V5).

V5

		Frequency	Percent	Valid Percent	Cumulative Percent
Valid	N	13	33.3	33.3	33.3
	A	23	59.0	59.0	92.3
	SA	3	7.7	7.7	100.0
	Total	39	100.0	100.0	

Table 1.6 : Frequency distribution of the response for the statement "**Usage of Integrated farming methods has increased my agricultural yield as compared to single cropping methods. "(V6).**

V6

		Frequency	Percent	Valid Percent	Cumulative Percent
Valid	SD	2	5.1	5.1	5.1
	D	1	2.6	2.6	7.7
	N	4	10.3	10.3	17.9
	A	24	61.5	61.5	79.5
	SA	8	20.5	20.5	100.0
	Total	39	100.0	100.0	

Table 1.7 : Frequency distribution of the response for the statement "**I have an overall satisfaction from the Integrated farming methods and its benefits "(V7).**

V7

		Frequency	Percent	Valid Percent	Cumulative Percent
Valid	N	3	7.7	7.7	7.7
	A	32	82.1	82.1	89.7
	SA	4	10.3	10.3	100.0
	Total	39	100.0	100.0	

Table 1.8 : Frequency distribution of the response for the statement "**After started using Integrated farming methods in my agricultural field, I feel economically empowered "(V8).**

V8

		Frequency	Percent	Valid Percent	Cumulative Percent
Valid	SD	1	2.6	2.6	2.6
	D	1	2.6	2.6	5.1
	N	3	7.7	7.7	12.8
	A	21	53.8	53.8	66.7
	SA	13	33.3	33.3	100.0
	Total	39	100.0	100.0	

Table 1.9 : Frequency distribution of the response for the statement **"After adoption of Integrated farming methods in my agricultural fields, I Feel my agricultural expenses have come down as compared to earlier agricultural practices. "(V9).**

V9

		Frequency	Percent	Valid Percent	Cumulative Percent
Valid	D	2	5.1	5.1	5.1
	N	7	17.9	17.9	23.1
	A	25	64.1	64.1	87.2
	SA	5	12.8	12.8	100.0
	Total	39	100.0	100.0	

Table 1.10 : Frequency distribution of the response for the statement **"I recommend the usage of Integrated farming methods to other farmer friends and relatives"(V10).**

V10

		Frequency	Percent	Valid Percent	Cumulative Percent
Valid	N	6	15.4	15.4	15.4
	A	20	51.3	51.3	66.7
	SA	13	33.3	33.3	100.0
	Total	39	100.0	100.0	

H0: Integrated farming system has a positive effect on farmers agriculture income.

Testing of hypothesis is carried out as, If the Z score is large which means the P-value is small in such case null hypothesis will be rejected. Otherwise null hypothesis is accepted. The hypothesis tested in this work using one Z- test is as follows.

Table 2.1: Z-test for the obtained primary data from respondents.

	N	Minimum	Maximum	Mean	Std. Deviation	Variance
	Statistic	Statistic	Statistic	Statistic	Statistic	Statistic
V1	39	1	5	3.85	.844	.713
V2	39	1	5	4.05	.826	.682
V3	39	1	5	3.95	.972	.945
V4	39	1	5	3.54	.884	.781
V5	39	3	5	3.74	.595	.354
V6	39	1	5	3.90	.940	.884
V7	39	3	5	4.03	.428	.184
V8	39	1	5	4.13	.864	.746
V9	39	2	5	3.85	.709	.502
V10	39	3	5	4.18	.683	.467
Z score (V1)	39	-3.37171	1.36691	0.00	1.00000000	1.000
Z score V2)	39	-3.69612	1.14921	0.00	1.00000000	1.000
Z score (V3)	39	-3.03384	1.08163	0.00	1.00000000	1.000
Z score (V4)	39	-2.87171	1.65341	0.00	1.00000000	1.000
Z score (V5)	39	-1.25052	2.11295	0.00	1.00000000	1.000
Z score (V6)	39	-3.08178	1.17271	0.00	1.00000000	1.000
Z score (V7)	39	-2.39406	2.27436	0.00	1.00000000	1.000
Z score (V8)	39	-3.62111	1.00916	0.00	1.00000000	1.000
Z score (V9)	39	-2.60559	1.62849	0.00	1.00000000	1.000
Z score (V10)	39	-1.72609	1.20076	0.00	1.00000000	1.000
Valid N (list wise)	39					

From table 2.1, we can observe the z- test statistics for all ten variables. All the ten variables have an observed mean higher than the Z mean. In other words Z values are significantly smaller than the observed values. Hence the Null hypothesis is accepted (**H0: Integrated farming system has a positive effect on farmer's agriculture income.**) and alternative hypothesis (**H1:**

Integrated farming system has no effect on farmers agriculture income.) will be rejected on the bases of observed values from the table 2.1.

FINDINGS

1. During the study it was observed that, farmers are more comfortable with integrated farming methods due to continues income all over the year.
2. Farmers have also expressed that, some time in a year they are facing price variability for their products in the market.
3. Farmers are also expressed the fact that , availability of water all over the year is a major constraint and decision factor for better yield in their farm land.

FUTURE SCOPE:

In this study the concentration was given to the farmers of Davangere district who are practicing the integrated farming. The future study can be extended to other geographical locations. Further the limited variables and samples are considered during the primary data collection due to the time constraint. It can be studied with more variables and sample size for the future study.

REFERENCES:

1. K.Ponnusamy and M. Kousalya Devi, Impact of integrated Farming System Approach on Doubling Farmers income, Agriculture Economic Research review volume 30, 2017 pp233-240.
2. Desh A.K and others, Empirical Proof on benefits of Integrated Farming system in Smallholder Farms in Odisha, Current Agricultural research journal, ISSN 2321-9971.
3. M. AkhtarHossain and others , Impact Assessment of integrated Farming System at FSRD Site,Goyeshpur, Pabna, Bangladesh. Asian Journal of Plant Science, Volume 2(2):167-170, 2003.
4. M Ansar and Fathurrahman ,Sustanable integrated farming system: A solution for national food security and sovereignty, Indonesia.
5. Arun K Sharma 2006. A hand book of organic farming Agrobios Jodhpur.

Abbreviations: SA-Strongly agree.   A- Agree.   N-Neutral. D-Disagree. SD-Strongly Disagree. Please mark your choice.

S.N		SD	D	N	A	S
1	I am aware of Integrated farming methods uses and I have used at least once in my farm land.	☐	☐	☐	☐	☐
2	I am happy with the performance of Integrated farming methods in my agriculture farm land.	☐	☐	☐	☐	☐
3	I am comfortable with adoption of Integrated farming methods in my agricultural field.	☐	☐	☐	☐	☐
4	Due to adoption of Integrated farming methods in my agricultural field, I avoid the usage of inorganic fertilizers.	☐	☐	☐	☐	☐
5	Adoption of Integrated farming methods has brought down the fertilizer cost in my agricultural field.	☐	☐	☐	☐	☐
6	Usage of Integrated farming methods has increased my agricultural yield as compared to single cropping methods.	☐	☐	☐	☐	☐
7	I have an overall satisfaction from the Integrated farming methods and its benefits.	☐	☐	☐	☐	☐
8	After started using Integrated farming methods in my agricultural field, I feel economically empowered.	☐	☐	☐	☐	☐
9	After adoption of Integrated farming methods in my agricultural fields, I Feel my agricultural expenses have come down as compared to earlier agricultural practices.	☐	☐	☐	☐	☐
10	I recommend the usage of Integrated farming methods to other farmer friends and relatives.	☐	☐	☐	☐	☐

YOUR KNOWLEDGE HAS VALUE

- We will publish your bachelor's and master's thesis, essays and papers

- Your own eBook and book - sold worldwide in all relevant shops

- Earn money with each sale

Upload your text at www.GRIN.com and publish for free